Where do you live? In a house? On a boat? In a mud hut?

In the city, some people live in high-rise apartments.

Some people live in cottages, like this one.

Some people live on canal boats or house boats.

In poorer parts of the world, many people live in slums.

In the Arctic, people live in igloos made out of ice and snow.

Some people live out of the city and in the country. They may live on a farm. Where do you want to live?